Rodrigo Medeiros

Automated Feeder for Rainwater Disinfection

Rodrigo Medeiros

Automated Feeder for Rainwater Disinfection

A low-cost automated dosing prototype

ScienciaScripts

Imprint

Any brand names and product names mentioned in this book are subject to trademark, brand or patent protection and are trademarks or registered trademarks of their respective holders. The use of brand names, product names, common names, trade names, product descriptions etc. even without a particular marking in this work is in no way to be construed to mean that such names may be regarded as unrestricted in respect of trademark and brand protection legislation and could thus be used by anyone.

Cover image: www.ingimage.com

This book is a translation from the original published under ISBN 978-613-9-65323-2.

Publisher:
Sciencia Scripts
is a trademark of
Dodo Books Indian Ocean Ltd. and OmniScriptum S.R.L publishing group

120 High Road, East Finchley, London, N2 9ED, United Kingdom
Str. Armeneasca 28/1, office 1, Chisinau MD-2012, Republic of Moldova, Europe
Printed at: see last page
ISBN: 978-620-7-86692-2

Dedication

I dedicate this book to my family, especially my parents, Gladstone and Mágna, who have supported and encouraged my academic and professional growth, and to my sister Railla.

To Professor Dr Márcio Valério de Araújo, for his patience, trust and ideas that made this book possible. For encouraging research and the development of prototypes, ever since he taught the subject of Measurement and Instrumentation Systems on the Mechanical Engineering course.

The Federal University of Rio Grande do Norte (UFRN), the staff and professors who make up the teaching staff and the management of the Mechanical Engineering course, for learning and contributing to my academic and professional development.

Summary

CHAPTER 1   3

CHAPTER 2   6

CHAPTER 3   26

CHAPTER 4   42

CHAPTER 5   50

CHAPTER 1

Introduction

Water is an essential natural resource for life on planet earth. It has immense economic, environmental and social value and is vital to the survival of man and our world's ecosystems. It is possibly the only natural resource that concerns all aspects of human civilisation, from agricultural and industrial development to the cultural and religious values ingrained in society.

Water is a finite and vulnerable resource that can represent an obstacle to a country's socio-economic development and the quality of life of individuals. There is an intrinsic relationship between access to good quality water, adequate sanitation infrastructure and human health (Philippi Jr A. et AL, 2005).

In this way, we realise the need to use alternative techniques to make use of water. In their respective works, both Lemos (2016) and Pordeus (2016) explain how rainwater and water from washing machine rinses can be used.

Harnessing rainwater for non-potable consumption has been used in several countries for years. This technology has been growing and emphasising water conservation, because as well as saving drinking water, it helps to prevent floods caused by torrential rain in large cities, where the surface has become impermeable, preventing water infiltration, showing itself as a possible alternative to be adopted (Miura, Silva and Lima, 2014).

Hespanhol (2015) explains that rainwater can contain coliforms, heavy metals, petrol and alcohol residues, among other dangers, which is why it must be properly treated before use.

Among the possible diseases that can be transmitted through contact between contaminated water and the skin, leptospirosis, caused by a bacterium called *Leptospirainterrogans,* and *schistosomiasis*, also popularly known as **water belly or snail disease, caused by the parasite** *Schistosoma,* are pathogens that can contaminate reused water.

According to the ABNT NBR 15527 standard, to reduce the risk of these types of pathogens, part of the water collected by the roof should be discarded, but this approach only reduces the risk of contamination, but does not eliminate it completely. For this, appropriate disinfection is recommended.

Based on this information, it is necessary to properly treat the water used for its intended purpose, as this will help to conserve water and prevent the transmission of diseases, thereby protecting human health and the environment.

Water treatment is summarised as a set of physical, chemical and biological procedures that are applied to water in order to remove contaminants, making the water suitable for use. In physical-chemical processes, filtering methods are applied in order to eliminate solid particles and start removing pathogenic microorganisms, and techniques for adding chemical products in appropriate proportions to completely disinfect the water.

There are various techniques for disinfecting water, such as boiling, adding chlorine ($Cl2$), chlorine dioxide, ozone ($O3$), iodine, silver salts, exposure to ultraviolet radiation and others. However, for the purpose of non-potable consumption, chlorination has the best advantages for disinfecting water.

Chlorination is the process of adding chlorine to water as a purification method. Chlorine is a highly efficient disinfectant, capable of eliminating pathogens such as bacteria, viruses and protozoa. Initially, chlorine was used as a disinfection agent to control epidemics; in 1902, it was used continuously to disinfect drinking water sources in Belgium. From 1908 to 1918, chlorination with small amounts of chlorine in the water was definitively initiated (MAYER, 1994).

Based on this information and knowing the lack of low-cost automatic systems for acquiring the components involved in disinfecting rainwater using 2.5% sodium hypochlorite or bleach as it is popularly known, this book sets out to develop a system capable of making the appropriate dosage for the level of reused water reserved for non-potable use.

In the case of rainwater utilisation systems, there are various ways of adding sodium hypochlorite, most of them by hand. However, the correct dosage of the products must be taken into account so as not to add too much or too little. Automated treatment systems can be used to facilitate the biological treatment of rainwater.

These automatic systems are capable of automatically controlling and providing tasks in an intelligent way, pointing out the advantages of operating autonomously, with little human intervention, they are capable of providing consistent and repeatable results, reducing the waste of raw materials as they work accurately in real time.

The automatic system developed used an Arduino UNO microcontroller, in which the data required for its exact execution is supplied by two level sensors, a peristaltic pump and a LED display to show this information.

The sodium hypochlorite level sensor sends an analogue signal to the microcontroller, which, on receiving it, converts it into a digital signal and sends it to the display, informing the operator of the amount of disinfectant available. At the same time, the rainwater sensor provides the Arduino with the discrete value of the water level in the storage tank. Based on this information, the controller triggers the peristaltic pump to start disinfecting.

This book presents the development of a low-cost automated closed-loop dosing system to disinfect rainwater for non-potable consumption using sodium hypochlorite as a disinfectant chemical, with the aim of preventing diseases transmitted by skin contact with contaminated water.

5

CHAPTER 2

Bibliographical review

There are a number of scientific papers on the subject of water treatment, many of which describe treatment methods as well as the chemical products used to disinfect rainwater for non-potable purposes.

According to the ANA/FIESP & SindusCon/SP manual (2005), systems consisting of simple sedimentation units, simple filtration and disinfection with solid chlorine or ultraviolet light are used to treat rainwater for non-potable purposes, while more complex systems can also be used to provide a higher level of quality.

Nakada (2008) presented the use of corn starch as a coagulant in cyclic filtration on a laboratory scale, obtaining satisfactory results.

In the study carried out by Barcelos & Felizzato (2005) on the use of atmospheric water for non-potable purposes, a filter made from an iron barrel was used. The barrel was 3.66 metres long and 1.33 metres in diameter. It was filled with sand and gravel to retain impurities.

However, most of these processes require the user to spend extra time maintaining the system, the availability of the disinfectant product on the market, since chlorine in solid form can only be found in specialised swimming pool stores and large supermarkets, and human exposure to the chemical.

2.1 Water demand.

Water demand refers to the volume that must enter the production and distribution system in order to meet consumer needs, losses and waste.

At the 6th World Water Forum in Marseille, France, issues such as population growth, increased demand for energy, food and climate pressures were presented as contributing factors to the growing demand for water. The dispute over water resources demands ever greater attention. In 40 years, demand is expected to grow by more than 50 per cent. Meanwhile, the planet's water resources are being contaminated. This is what the United Nations (UN, 2013) report says.

According to a UN report, the world's demand for water will grow by 55 per cent by 2050. Meanwhile, population growth over the next 40 years is estimated at two to three billion people. Tucci, who holds a PhD in Water Resources from *Colorado State University* and is a professor at the Hydraulic Research Institute of the **Federal University of Rio Grande do Sul (UFRGS), explains the problem. "There are** two risks: the risk of scarcity due to increased demand (more users and demand) and scarcity of quality due to contamination of the **available** water." Figure 1 shows an outlook for the availability of fresh water over the next few years.

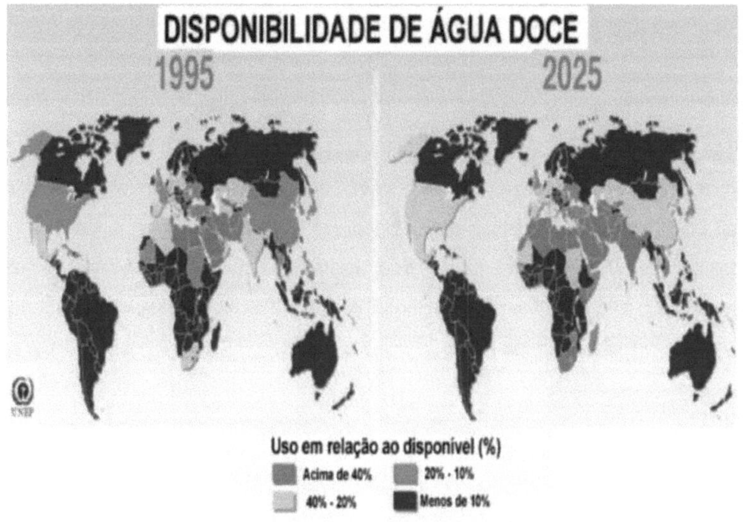

Figure 1 - Availability of fresh water.

Source: World Meteotological Organisation (WMO).

According to Professor Tucci, demand is growing not only because of population growth, but also because of changes in habits, rising incomes and **other factors. "It's not the water that can run out, but the increase in demand** that makes the same quantity be disputed by a greater number of users, as well as **the reduction in availability due to contamination" (Tucci,** 2013).

The United Nations Educational, Scientific and Cultural Organisation (UNESCO) estimates that by 2025, two thirds of the world's population will be affected in some way by a lack of drinking water. The United Nations World Report on the Development of Water Resources reinforces the desire to improve water management and reduce waste in order to meet the demands of an increasingly large and urban world population and the agricultural and industrial sectors.

For Tucci, a doctor in water resources (Tucci, 2013), the population is used to paying little and wasting water, without caring to see a river or stream contaminated.

2.2 Use of reusable water sources.

Although it may seem like a new need, the search for an alternative source of water has been going on for thousands of years, for both potable and non-potable purposes.

Alternative water sources are proving to be a viable way of meeting water demand. According to the manual Conservation and Reuse of Water in Buildings (ANA/FIESP/SindusCon/SP, 2005), alternative water sources are all those that are not under concession from public bodies or that are not charged for their use.

Recent years have seen the development of new technologies for managing water resources. This has led to new expansions in the use of rainwater utilisation techniques, both in regions where they were already used and in places where they were unknown (PETRY & BOERIU, 2000 cited by MAY, 2004).

Utilising rainwater as an alternative source is a simple and inexpensive way of preserving drinking water, given that the purchase costs of the components involved are low. It is feasible in regions where rainfall is generous in terms of quantity and distribution throughout the year.Figure 2 shows rainwater harvesting in a home. This collection model is very practical and easy to incorporate into any household, as can be seen from the image.

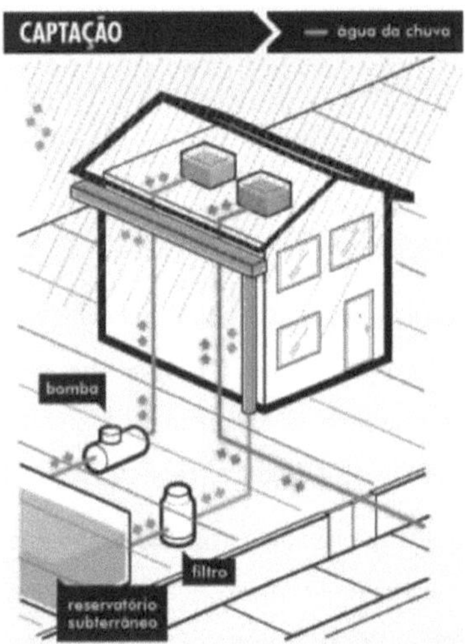

Figure 2 - Rainwater harvesting.
Source: www.meioambiente.culturamix.com

The practice of rainwater harvesting is common in various regions, especially in arid and semi-arid areas of the world where scarcity is felt most acutely. In areas with low and irregularly distributed rainfall, rainwater harvesting has been used to supply water for human consumption (GNADLINGER, 2009; QIANG, 2009; TAVARES, 2009).

Rainwater is currently being utilised in several countries on different continents, many of which offer benefits for the construction of rainwater harvesting and storage systems, such as the United States, Germany and Japan. In these countries, for example, the process of rainwater harvesting began with the aim of retaining rainwater as a preventative measure to combat urban flooding. However, over time, the use of water has gained ground due to the risk of scarcity and also to promote the recharge of the subsoil, which is the main source of water supply in these countries (GROUP RAINDROPS, 2002).

In Germany, rainwater, which is always used for non-potable purposes, has been used for flushing toilets, irrigating gardens, washing machines, commercial and industrial use since the 1980s (TOMAZ, 2010).

Gelt (2002) and Moffa (1996) apud May (2004) describe that in the United States rainwater harvesting is used for flushing toilets, evaporative cooling, washing vehicles and irrigating gardens and vegetable plots.

In Brazil, the utilisation of rainwater has been applied in some semi-arid regions. There are projects that promote the implementation of systems aimed at harnessing this water, such as the federal government's One Million Cisterns Project (P1MC), which aims to guarantee a supply of water to the poorest population during periods of drought (CARVALHO, 207).

The reuse of water from washing machine *rinses* is also an alternative source, as suggested by the article *Home Use ofGreywater, Rainwater Conserves Water- and May Save Some Money* (1993) available on the University of Arizona website.

Figure 3 shows a schematic illustration of an automated washing machine water reuse system, incorporated into a home. After reuse, the water is redirected to the top box by a pump and then distributed for non-potable purposes.

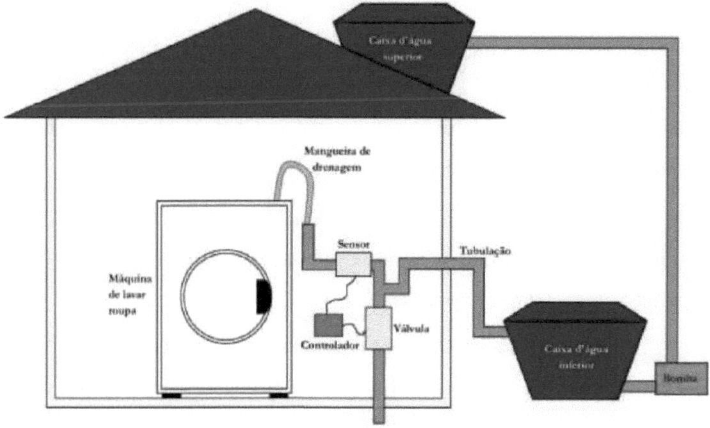

Figure 3 - Schematic drawing of the washing machine water reuse system.

Source: Pordeus (2016).

According to (Pereira, 2013), the reuse of water from air conditioners installed in homes and condominiums is another alternative source. The drops that come out of the refrigerator can add up to several litres at the end of the day, allowing it to be reused in sustainable practices. Figure 4 shows a reservoir for utilising condensed air-conditioning water.

Figure 4 - Air-conditioning condensate tank.
Source: www.jie.itaipu.gov.br

2.3 Need for treatment.

During the hydrological cycle, water is exposed to contaminants from humans and animals. According to Cavinatto (1992), since ancient times man has learnt intuitively that water polluted by waste could transmit diseases.

According to Pinheiro (2016), waterborne diseases are predominantly caused by microorganisms exposed in freshwater reservoirs, usually after contamination by human or animal faeces. Transmission of the infectious agent through water can occur through skin contact during bathing, ingestion or aspiration of germs present in the water.

According to the World Health Organisation (WHO) (1999), in Latin America and the Caribbean, water-borne diseases caused by the contamination of water resources used for human consumption have had a major impact on the population.

According to the Brazilian Association of Sanitary Engineering - ABES (1994), around 65 per cent of hospital admissions in Brazil are the result of water-borne diseases, worsening the public health situation and increasing infant mortality and morbidity rates.

According to the World Health Organisation (WHO), a large proportion of all the diseases that spread in developing countries come from poor quality water. Contaminated water can harm people's health in the following situations:

- Through direct ingestion;

- Food intake;

- For use in personal hygiene and leisure;

- In agriculture;

- In industry.

Water-related diseases can be grouped according to Table 1 below.

Table 1 - Water-borne diseases.

Group of diseases	Forms of transmission	Main diseases	Forms of prevention
Transmitted by the faeco-oral route	0 pathogenic organism is ingested	Diarrhoea and dysentery; cholera; giardiasis; amoebiasis; ascariasis (roundworm)...	- Protect and treat water supplies and avoid using contaminated sources...
Controlled by cleaning with water (associated with insufficient water supply)	Lack of water and poor personal hygiene create favourable conditions for its spread	Skin and eye infections, such as trachoma and lice-related typhus, and scabies	- Provide adequate water and promote personal and domestic hygiene
Associated with water (part of the life cycle of the infectious agent occurs in an aquatic animal)	0 pathogen penetrates through the skin or is ingested	Schistosomiasis.	- Avoid contact with infected water; - Protect water sources.
Transmitted by water-borne vectors	Diseases are spread by insects that hatch in or near water.	Malaria; yellow fever; dengue fever; filariasis (elephantiasis)	- Fight the insects that transmit them; - Eliminate conditions that may favour breeding sites.

Source: Barros et al. (1995).

Still on the subject of diseases, the municipal health department of the city of São Paulo warns against contamination by leptospirosis. This disease can be transmitted through direct contact between humans and the urine of infected animals or indirectly, in the most common case, through contact with contaminated water or soil.

Based on these findings and information, it became clear that there was a need to provide appropriate treatment in order to eliminate or reduce the concentration of these pathogens.

2.4 Water disinfection.

The purpose of disinfection is to destroy or inactivate pathogenic organisms or other undesirable organisms capable of producing disease or not. This process consists of applying a physical, chemical or biological agent, in different techniques that are convenient for extinguishing infectious micro-organisms, ensuring and protecting humans from infections that come from contaminated water.When differentiating water disinfection processes, we can divide them into two groups, disinfection for potable and non-potable purposes. The main dividing factor is the intensification of these processes.

2.4.1 Drinking purposes.

2.4.1.1 Boiling.

Heat is a physical agent used to disinfect water. The Ministry of Health recommends filtering and boiling water for three minutes before drinking it. This act helps exterminate disease-causing microorganisms and is a common practice to make water drinkable in less favoured regions.

2.4.1.2 Ultraviolet (UV) rays.

Ultraviolet (UV) radiation was discovered as a germicide in 1877 by Downes and Blunt and has been used as a disinfectant in public water supplies since 1900 (KOLLER, quoted by CAMPOS et al., 1991).

Figure 5 shows a UV unit for treating water using ultraviolet rays. The diagram shows the water entering from one side and the fluid being exposed to ultraviolet radiation for a certain time before being removed.

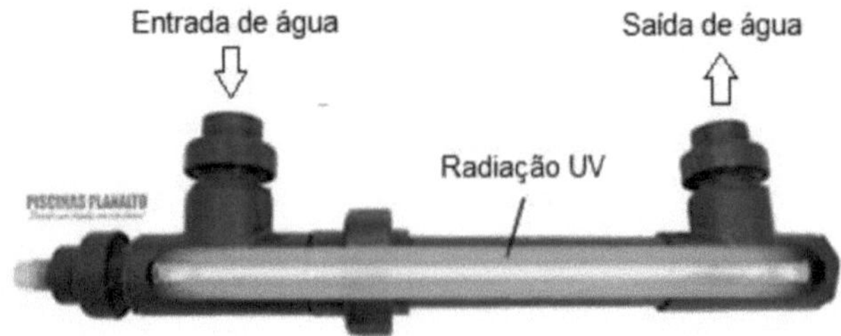

Figure 5 - UV unit.
Source: planalto pools (2016) modified by the author

This process consists of transferring electromagnetic energy from low-pressure mercury vapour lamps to the genetic material (DNA and RNA) of pathogenic microorganisms, destroying their cells' ability to reproduce.

According to Campos (apud CAMPOS et al, 1991), when radiation passes through the cell wall and reaches the DNA, it generates 12-pyrimidine dimers from the same chromosome strand, which prevents the normal duplication of microorganisms.

2.4.1.3Ozone .

Ozone was designated as a germicide in 1882, when Ohlmuller realised that this gas could render cholera and typhus bacteria inactive. In 1886, it was introduced in France for water treatment. In 1891, in Germany, using a pilot-scale plant, Frölich analysed and demonstrated the propensity and efficiency of ozone in water treatment. It was used for the first time in 1893 in Oudshoom, Holland. In Nice, France, in 1906, it was used as part of the treatment of drinking water (ELLIS, apud GRASSI and JARDIM, 1993).

Ozone as an oxidiser and disinfectant is exceptionally effective in eliminating pathogenic microorganisms. According to the EPA (1999a), the mechanisms of disinfection using ozone occur in various ways, such as the extinction of the cell wall of micro-organisms providing the shedding of the internal components of the cell or the reaction with ozone decomposition by-products, which damage the constituents of nucleic acids (purines and pyrimidines).

Figure 6 shows a simple ozone water treatment system for swimming pools. The water to be treated first passes through a filter, then a pump, and finally an ozone generator is introduced into the water line to disinfect it.

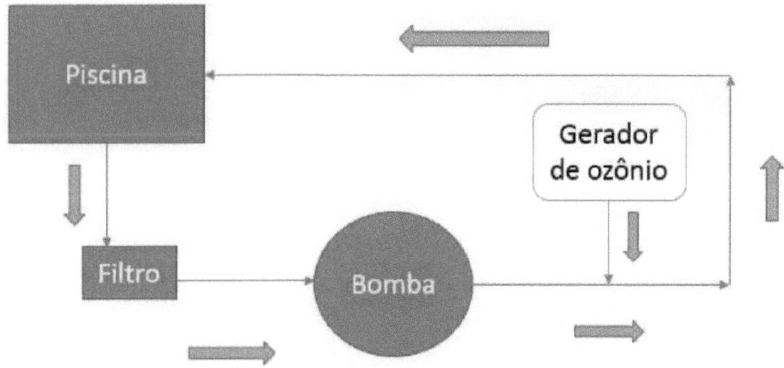

Figure 6 - Ozone pool water treatment system.
Source: prepared by the author

2.4.1.4Iodine .

Iodine has been used to guarantee drinking water since the 1940s, when they developed a formulation of pills used by troops in the field (Chang S, Morris J. 1953). Figure 7 shows a box of iodine pills used to disinfect water.

Figure 7 - Box of iodine pills
Source: prepared by the author

Iodine is a halogen, like chlorine, which exerts a biocidal effect through its chemical property as a strong oxidiser. The active species of disinfectant is the iodine component (Gottardi W. Philadelphia:Lea&Febiger, 1991;152-167. White G 1992.).

Some restrictions should be considered when using iodine as a disinfectant for drinking water. Due to the risk of some people being allergic to the substance, pregnant women, women over 50 and people with thyroid problems should consult their doctors before using these water disinfection processes, the World Health Organisation (WHO) and the US Environmental Protection recommends that iodine be used on a short-term basis, only in emergency cases.

2.4.1.5 Chlorine dioxide (CIO_2).

Chlorine dioxide (CIO_2) was discovered in 1811 by SIR HUMPHREY DAVY, who **called it** *"teegreen-eelowgaeechhionee".* **DAVY produced the gas by acidifying** potassium chlorate with sulphuric acid, and the first reference in the literature was by Millon, who obtained the green-yellow gas by acidifying potassium chlorate with hydrochloric acid. He absorbed the gas in an alkaline solution and obtained chlorite (and chlorate). Millon's Gas, as it was called, was not identified as containing chlorine dioxide until it was identified in 1881 (AIETA et al, 1986).

TAYLOR et al. (apud AIETA et al., 1986), announced the discovery of a new commercial chemical product available for bleaching, with superior characteristics to those used until then: sodium chlorite. They also discussed the release of chlorine dioxide under acidification, i.e. reaction with chlorine. Since then, the use of chlorine dioxide for water treatment has been made possible by the commercial availability of sodium chlorite.

Chlorine dioxide is a fast-acting disinfectant, equal to or superior to chlorine in inactivating bacteria and viruses. This gas is also effective in destroying pathogenic protozoan cysts (NARKIS,1996, AIETA et al., 1986).

2.4.1.6 Chlorine.

Chlorine disinfection is one of the most common practices worldwide for disinfecting water intended for public supply and for treated effluents.

According to SIGHIERI (1974), it was discovered by chance by CARL WILHELM SCHEELE when an experiment with muriatic acid and manganese dioxide gave off a greenish-yellow gas called **oxygenated** muriatic acid. **It inherited the name "chloro" in 1810, when SIR HUMPHREY DAVY** proved that this gas was a chemical element and in 1888, in Germany, it was made commercially in the form of liquid chlorine.

At first, its application as a disinfection agent was aimed at controlling epidemics; in 1902, it was used uninterruptedly to disinfect drinking water in Belgium. From 1908 to 1918, chlorination with small amounts of chlorine in the water was definitively introduced (MAYER, 1994). Not long ago, chlorination was essentially the most widely used option in the United States, some European countries and most developing countries, due to its low cost and efficiency in killing pathogenic microorganisms.

2.4.1.7 Bleach.

Bleach is a commercial product obtained by diluting sodium hypochlorite in water, which is stabilised by adding sodium chloride. According to manufacturers' data expressed on the labels of this product, the active chlorine content is generally around 2.0 to 2.5%. According to Thayer (2007), this substance is called bleach and has a high germicidal power, which is why it is used by housewives for washing clothes and disinfecting many vegetables.

According to data from the Rondônia state government portal (2016), bleach in a proportion of 1 litre for every 1,000 litres of water stored in the reservoir can be applied as a disinfectant to combat leptospirosis.

Plants can tolerate chlorine levels below 70 mg/L without any harmful effects, as can be seen in Table 2. According to Ribeiro (1999), plants' ability to absorb nutrients depends on the amount and types of substances found in the soil and irrigation water.

Table 2 - Potential problems related to water quality.

Features	Damage level		
	None	Medium	Severo
pH	5,5-7,0	< 5.5 or > 7.0	<4.5 or > 8.0
C.E (dS/m)	0,50,75	0,75-3,0	> 3,0
Total soluble solids (mg/L)	325 - 480	480-1920	> 1920
Bicarbonates (mg/L)	<40	40-180	> 180
Sodium (mg/L)	< 70	70-180	> 180
Calcium (mg/L)	20-100	100-200	>200
Magnesium (mg/L)	<63	>63	
RAS	< 3,0	3,0-6,0	>6,0
Boron (mg/L)	< 0,5	0,5-2,0	>2,0
Chlorine (mg/L)	< 70	70 - 300	> 300
Fluoride (mg/L)	< 0,25	0,25-1,0	> 1,0
Iron (mg/L)	<0,2	0,2 -0,4	> 0,4
Nitrogen (mg/L)	< 5,0	5,0 - 30,0	> 30

Source: Ribeiro 1999.

In view of its ability to combat leptospirosis and other diseases, and the fact that it can also be used for watering gardens if dosed in appropriate proportions, as well as being easily found and purchased anywhere at a relatively low cost, sodium hypochlorite (bleach) is the disinfectant chosen to operate in the automated system developed in this book.

2.4.2 Non-potable purposes.

Collecting water for non-potable purposes does not require much purification, although a certain degree of filtration is often necessary. For elementary treatment, natural sedimentation, simple filtration and chlorination processes can be used. If rainwater is to be used for human consumption, more complex treatments such as ultraviolet disinfection or reverse osmosis are recommended (MAY & PRADO, 2004).

2.4.2.1 Filtration.

There is a range of filters on the market that can be incorporated into rainwater harvesting projects. Their function is to retain solid particles, because according to Annecchini (2005), it is necessary to prevent coarse materials such as leaves, twigs and sticks from reaching the final water storage reservoir, as they could lead to a reduction in water quality due to decomposition processes. The filters are installed manually once and only need to be replaced over time, but regular maintenance is necessary to prevent them from clogging up, otherwise they will not be able to fulfil their purpose. Figure 8 shows a low-cost self-cleaning rainwater filter. Usually installed in homes with rainwater harvesting systems.

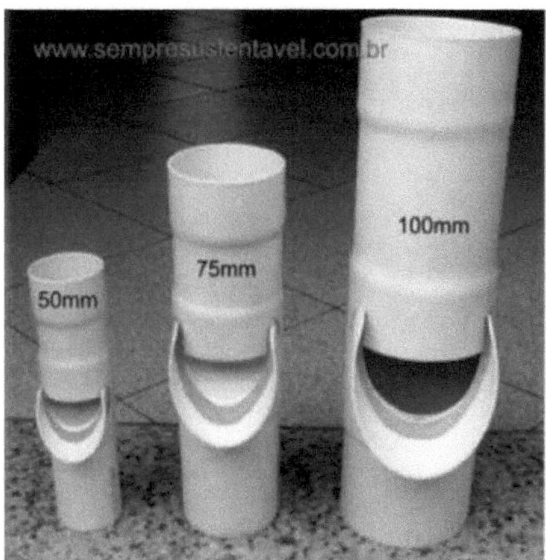

Figure 8 - Self-cleaning model filter.
Source: sempresustentavel.com.br

2.4.2.2 Sedimentation   .

Sedimentation is a unitary operation found in water treatment plants, based on the difference in density between water and suspended solids, resulting in the deposition and subsequent removal of these solids, and is one of the most common processes in water treatment (medeiros, 2014). Its mechanism consists of removing the suspended solids manually, causing labour for the operator and risk when coming into contact with some contaminated substances.

2.4.2.2 Chlorination.

The ABNT NBR 15.527/07 standard, shown in Table 3, proposes the use of chlorine compounds to disinfect rainwater for non-potable purposes. In this table we can observe the qualitative parameters of rainwater and the appropriate proportions of chlorine in order to achieve a satisfactory dosage to eliminate health risks.

Table 3 - Rainwater quality parameters for restrictive non-potable uses

Parameters	Analysis	Value
Total coliforms	Half-yearly	Absent in 100 ml

Thermotolerant coliforms	Half-yearly	Absent in 100 ml
Free residual chlorine [a]	Monthly	0.5 to 3.0 mg/L
Turbidity	Monthly	< 2.0 uT[b] , for less restrictive uses <5.0 uT
Apparent colour (if no dye is used, or before use)	Monthly	< 15 uH [c]
It must provide pH adjustment to protect the distribution networks, if necessary	Monthly	pH of 6.0 to 8.0 for carbon steel or galvanised pipes
[a] NOTE Other chlorine processes can be used, such as the application of ultraviolet rays and[b] uT is the unit of turbidity. [c] uH is the Hazen unit		and disinfection in addition to ozone application.

Source: ABNT NBR 15527.

There are several ways of chlorinating rainwater for non-potable purposes. The addition of the disinfectant additive can be carried out manually or automatically, using chlorine in liquid or solid form (granules or tablets). Figures 9, 10 and 11 below show some of the types of chlorinators used to disinfect water.

Figure 9 - Nautilus tablet chlorinator.

Source: www.nautilus.ind.br

This chlorinator is intended for use only with disinfectants in solid form, in the form of tablets or low-dissolution chlorine tablets (trichlor). Its capacity is up to 9 x 200 gram tablets or approximately 1,900 kg of smaller tablets and it has a valve to regulate the flow of water through the chlorinator.

Figure 10 - V1.5 220 P4 filter feeder.

Source: www.viafiltrosshop.com.br

This electronic chlorine doser provides automatic dosing without variations in the injection of the chemical, using the substance in a liquid state. It is normally used when high precision and an excessive working regime are required.

Figure 11 - Metering pump.
Source: www.emecbrasil.com.br

Among these and other chlorinators, those classified as using solid chlorine (tablets or granules) have some disadvantages compared to those that use the disinfectant in liquid form, since this product can only be purchased in specialised shops, such as swimming pool shops, and not all towns have this type of shop. It also costs more than bleach.

2.5 Dosing pumps.

The metering pump is a piece of mechanical equipment that carries out fractional transfer and rigorous distribution of fluids, i.e. it has precise control of the flow rate or volume pumped over time. As such, this device covers a wide range of applications in the many branches of industry that involve dosing, such as the introduction of reagents for pH control, the addition of chemical products for disinfection, the injection of yeast in brewing, among others.There are three types of metering pumps found on the market, between which their operating principles differ.

2.5.   1Diaphragm metering pump.

It is a positive displacement pump, usually reciprocating, in which the pressure increase is realised by the impulse of an elastic wall (membrane or diaphragm) which alternates the volume of the chamber, periodically increasing and decreasing it. Check valves, usually made of elastomer balls, control the movement of the fluid from the lower pressure zone to the higher pressure zone. Figure 12 shows a simple model of a diaphragm metering pump.

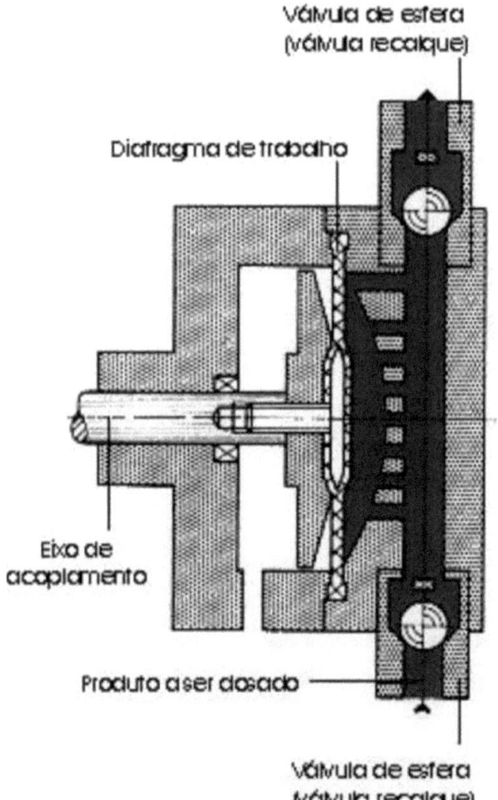

Figure 12 - Single diaphragm pump

Source: www.inovatronic.com.br

2.5.   2Peristaltic dosing pump.

20

The peristaltic pump is an excellent multifunctional option for applications in various areas, with low maintenance and excessive wear rates. Its main feature is that the pumped liquid does not come into contact with any part of the equipment, the fluid only comes into contact with the inner surface of the hose, so fluid with oxidising characteristics can be incorporated into the pump's operation as the internal parts are not damaged by the liquid and it is not contaminated.

In a peristaltic pump, the hose is properly positioned and fixed to the pumphead and is pressed by rollers around a rotor that moves in a circular motion. As the rollers rotate around the rotor, they compress and close the hose, creating the vacuum needed to move the fluid. After passing the roller, the hose returns to its nominal diameter due to the particular conformation of the material from which the hose is made. Figure 13 shows a peristaltic pump for chemical products.

Figura 13   - Peristaltic pump for chemical products.

Source: www.servproject.com

2.5.3 Gear metering pump.

A gear pump is a pump that creates a certain flow rate due to the constant meshing of two or more gear wheels. The two gears are housed in a casing, one of which (the driving gear) has a through shaft that transmits the power supplied by the motor. The other gear that does the meshing is called the driven gear. The constant disengagement of the teeth creates decompression in the suction chamber, causing the fluid to be sucked out of the reservoir. It is then driven peripherally through the wheel openings that form a closed chamber with the pump casing and side seals. Constant gearing expels the fluid from the shafts and forces it out of the pump. Figure 14 shows the working principle of a gear pump.

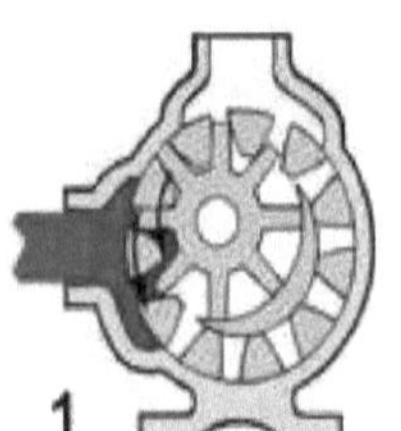

Figure 14 - Working principle of the gear pump.
Source: IFF (2016)

2.6 Sensors.

According to Accardi and Dodonov (Aput Almeida, 2009), sensors are devices that identify impulses, measure and monitor physical quantities and events (temperature, humidity, etc.), transforming them into a value that can be manipulated by computer systems. They provide the controllers with information about a given event, so that the controllers can send the appropriate commands to the actuators.

According to Fuentes (2005), sensors can be classified in two ways, according to the nature of the output signal: discrete sensors or analogue sensors. Discrete sensors are used to monitor the occurrence or non-occurrence of a particular event, and have two different output states, such as on or off. Analogue sensors are used to monitor a physical quantity over a continuous range of values set between minimum and maximum limits. There are various types of sensors for which their employability depends directly on the type of physical quantity to be analysed.

2.6.1 Level sensor.

"Level is one of the most common and widely used variables in industrial applications. Level measurement is defined as determining the **position of an interface between two media"** Bega et al. (2006). The level sensor fulfils this definition.

There are two types of level sensors: discrete and continuous. Discrete sensors are those that generate an output signal when they identify that the liquid is at a given height (float with level switch, metal electrode, etc.). Continuous level sensors are those that provide an output signal proportional to the level of the liquid (ultrasonic distance sensors, monitoring the weight of the liquid using a load cell, among others).

2.6.1.1 Metal electrode type level sensor.

The metal electrode level sensor is only suitable for electrically conductive liquids. It is based on two or more metal rods, situated in different horizontal positions, which are submerged in the fluid and powered by an electrical voltage source. As the level expands, the electrodes come into contact with the water, allowing the electric current to pass through and closing the circuit. In this way, it is possible to measure discrete levels in a reservoir. Figure 15 shows a simple schematic of an electrode-type level sensor.

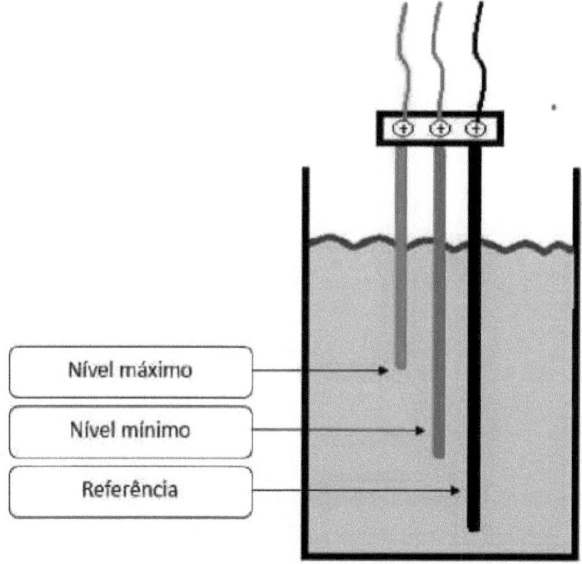

Figure 15 - Schematic of the metal rod submerged in water.

Source: www.embarcados.com.br.

2.6.1.2 Float-type level sensor.

The float sensor consists of a float element (float), which has a lower density than the liquid being measured, plus an element that is fixed in a certain position. As the fluid level rises, the float is pushed upwards by the thrust produced by the fluid. This displacement can cause the contact to open or close by mechanical action.

This gauge is often used to activate pumps. When the minimum fluid level is reached, the sensor detects this information and closes the pump drive contact so that the reservoir is filled. Later, when the level reaches its maximum capacity, the pump contact is opened, switching it off. Figure 16 shows a float switch with a pigtail.

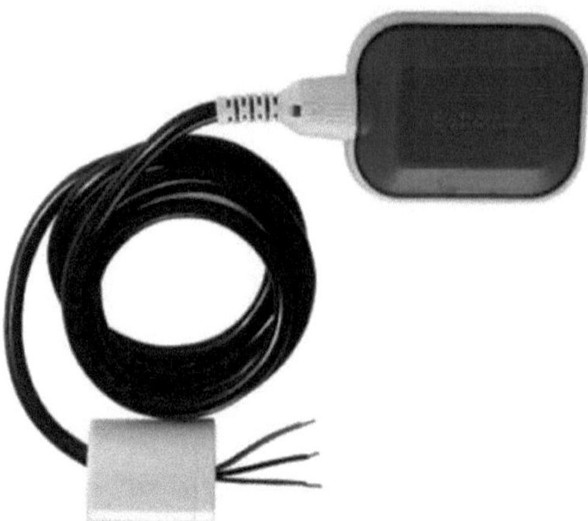

Figure 16 - UKY-2 Eletromar float switch with pigtail
Source: www.efacil.com.br

2.7 Automated systems.

For Rosário (2009) automation can be defined as:

> [...] a set of techniques for building active systems capable of acting with optimum efficiency by using information received from the environment in which they operate. Based on the information received, the system calculates the most appropriate corrective action, i.e. an automation system behaves like a human operator, using sensory information. It thinks about and executes the most appropriate action.

Ribeiro (2001) calls automation the idea of using electrical or mechanical power to drive some kind of machine. "It must add some kind of intelligence to the machine so that it performs its task more efficiently and with economic and safety advantages" **Ribeiro (2001).**

There are numerous advantages to automation, such as increased process efficiency, lower costs, improved quality standards, greater control and operational safety Rosário (2009). Another benefit is the capacity for interaction between several devices operating simultaneously.

According to Pinheiro (2004), an automated system is a group of devices that work together to assign tasks or produce a product or a family of products. Their use has been widespread, from applications with a simple float capable of controlling the level of a reservoir to digital systems in sophisticated aircraft.

These systems can be classified as open or closed loop. The main point of distinction between them is whether or not the control action is affected by the output signal.

In the closed loop, the output signal has a direct effect on the control action. In this type of system, the error signal, which corresponds to the difference between the reference and feedback values (which can be the output signal or a function of the output signal), is fed into the controller in order to reduce the error and maintain the system output at a certain value desired by the operator.

CHAPTER 3

Methodology

The automated disinfection system, based on liquid chlorine compound, for rainwater for non-potable purposes developed in this book, proposes, in its basic aspect, to automatically chlorinate rainwater in an appropriate proportion without wastage. The aim is to eradicate the pathogens present in this type of water, with as little human interaction as possible.

To this end, the model developed works as a closed-loop automated system. It consists of two sensors, a peristaltic pump, an Arduino UNO microcontroller, a reservoir for storing sodium hypochlorite and an LCD display (16x2) for visualising information on the actual rainwater level in the reservoir, as well as informing the need for a sanitary water supply.

After chlorination by the prototype, rainwater becomes suitable for non-potable use, such as washing pavements, cars, watering gardens, filling toilets and more. This contributes to rationalising drinking water consumption in a household when the use of this disinfected water is required.

Figure 17 shows a schematic drawing based on the connection of the components involved in the development of the automated system and its operating principle, showing the direction in which the information flows.

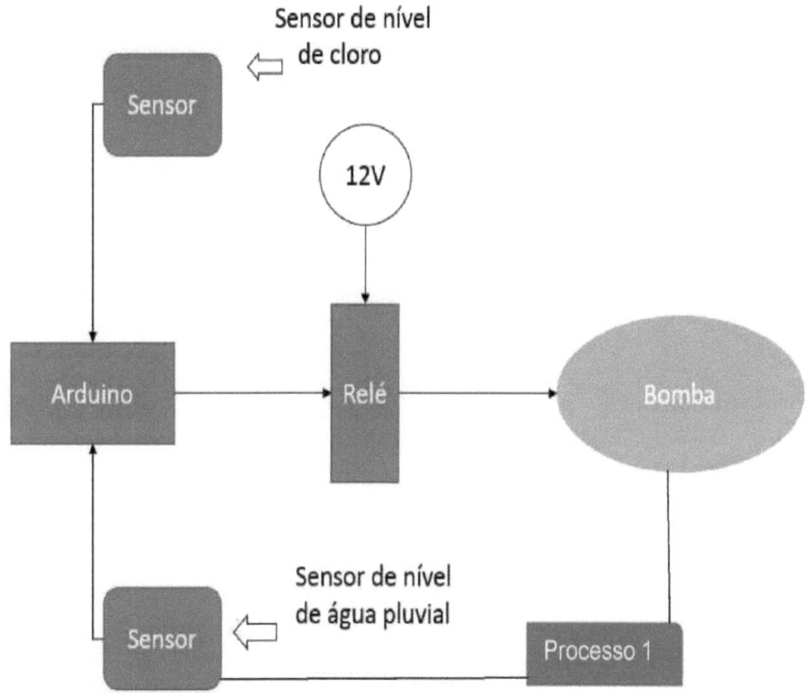

Figure 17 - Schematic drawing of the automated chlorine doser.
Source: prepared by the author

Once the rainwater has been collected by a catchment method, the device designed for this book must be switched on. When triggered, the chlorine level sensor immediately sends an analogue signal to the microcontroller, informing it of the presence or absence of sodium hypochlorite in its tank. This signal is then converted into digital by the Arduino itself, which, after processing, sends a command to the display so that it can show the operator messages alerting him to the replenishment of the disinfectant in his tank.

At the same time, when the rainwater meter sensor checks the level of rainwater in its reservoir, it informs the microcontroller of the discrete quantity of this fluid, after which the Arduino UNO receives and processes this data, sending information to be shown again on the display about the discrete quantity of water, while at the same time sending information to the peristaltic pump drive relay, closing or continuing with the electrical circuit open based on the information received. Using a hose, the peristaltic pump extracts sanitary water from its tank and directs it to the rainwater reservoir, chlorinating it accordingly.

After the last disinfection, the system must be switched off. It is essential that the user waits 30 minutes for the chlorine to perform its disinfectant action, after which the water is free to be used for non-potable purposes.

3.1 Chlorine tank

A simple, low-cost disinfectant tank was designed for this book, with a storage capacity greater than the volume needed to completely disinfect the rainwater tank, made from a material that does not act chemically with chlorine and that could be sealed to prevent contamination. This reservoir was divided into two segments, the main and secondary.

3.1.1 Main reservoir.

The main reservoir holds the largest volume of sodium hypochlorite stored, with a capacity of 2.1 litres. A plastic tank with a screw cap and a welded adapter with a 1-inch **water** tank ring were used to make the reservoir. Figure 18 shows an illustration of the assembly of the main tank.

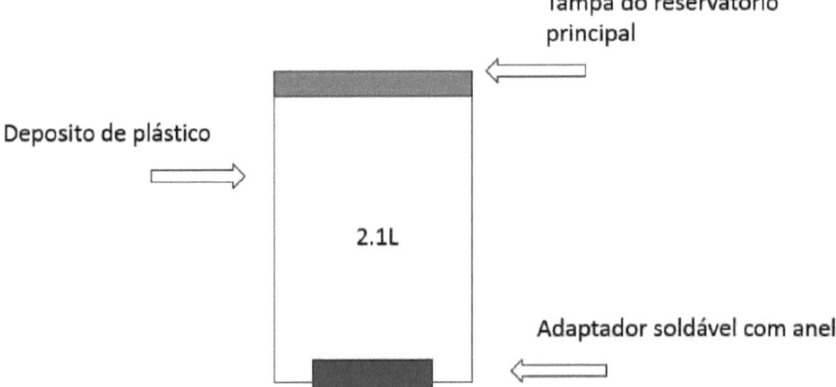

Figure - 18 Illustration of the main reservoir.
Source: prepared by the author

3.1.   2Secondary reservoir.

The secondary reservoir has been designed to contain the volume of a single dosage, 166ml. It consists of a short 1-inch weldable adapter, a 250mm PVC pipe with an external diameter of 1 inch and a 1-inch cap connected at the bottom and coupled to the peristaltic pump's suction hose, this segment being the body of the chlorine level sensor. Figure 19 shows an illustration of the secondary reservoir.

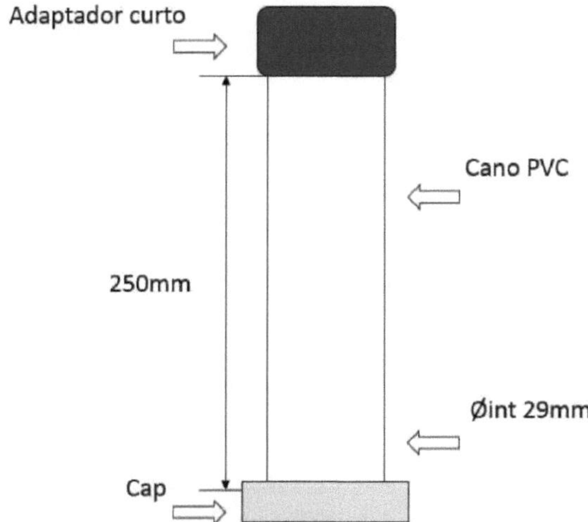

Figure 19 - Illustration of the secondary reservoir.
Source: prepared by the author

3.2 Sensor.

In this book, two sensors have been developed to detect the discrete level of liquids, but their operating principles are dissimilar.

3.2.1 Chlorine level sensor.

The chlorine level sensor was based on the action of a presence sensor, with simple performance and low component acquisition costs. It required a pair of 5mm infrared light-emitting diodes (LEDs) and a pair of infrared phototransistors. These components were installed in a piece of PVC pipe. Figure 20 shows a schematic drawing of the chlorine level sensor developed in this book.

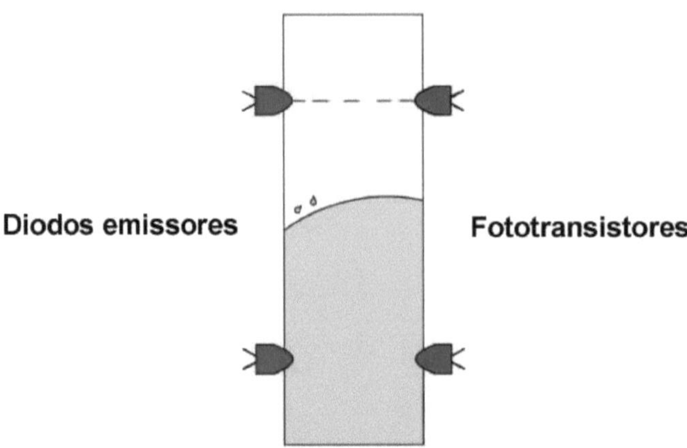

Figure 20 - Schematic drawing of the chlorine level sensor.
Source: prepared by the author

Each pair, emitter and receiver, was coupled so that they were facing each other, thus forming a retroflection-type presence sensor. The working principle of this sensor is as follows: when switched on, the emitter diode emits infrared light, which is not visible to the naked eye, directly to the phototransistor, which measures the intensity of the infrared rays received and sends the analogue signal generated to the microcontroller, which then converts it into a digital signal. When no infrared rays hit the receiver, Arduino reports a value of zero, while when the receiver receives as much light as possible, the microcontroller reports values close to 1023.

The upper end of the sensor was attached to the disinfectant tank and the lower end connected to the peristaltic pump via a hose in order to drain the fluid. When bleach is added to the tank, the liquid fills the entire chlorine level sensor pipe, so the emitting diodes are covered, consequently the emitted infrared rays are deflected or reach their respective phototransistors with little intensity, and so the Arduino displays values close to the 939 to 967 range.

When draining is carried out and therefore the diodes are not exposed to contact with the liquid, most of the infrared rays emitted are detected by the phototransistors, so the Arduino displays values in the range 1009 to 1018.

To build the sensor, care was taken to control the distances at which the diode pairs would be mounted, since incorrect positioning of the LEDs could interfere with the function the device was designed to perform. This sensor was developed with the aim of alerting the operator when the last dosage has been made and when the reservoir needs to be refilled with the chemical product via the display.

The volume of bleach estimated for a dosage was determined implicitly when the rainwater level sensor was manufactured, since the distance at which the electrodes were attached determines the volume of water that needs to be treated. The proportion of 1 litre of bleach per 1000 litres of water was strictly adhered to, because according to the Rondônia state government website, this amount of disinfectant is capable of eradicating the leptospira bacteria responsible for leptospirosis. We also know that this amount is less than the 70mg/L of chlorine taken up by vegetables without running any risk, which was also mentioned in the literature review.

Because seven electrodes are used in the manufacture of the rainwater sensor, logically, with this quantity we can measure 6 equal distance intervals and state that each of these intervals measures 166.67 L of water, because through a simple ratio between the maximum fluid capacity stored by the rainwater reservoir, which is exactly 1000 L, and the number of intervals available in the device, we arrive at this result. Therefore, using a simple rule of three, we conclude that the volume corresponding to one disinfection is 166.67 ml of disinfectant.

Using a measuring instrument, we found that the internal diameter of the pipe in which the LEDs were installed was 29mm, and using equations 1 and 2 to calculate the area of the circumference and volume of the cylinder, respectively, we came to the conclusion that these diode pairs must be situated at a distance of approximately 250mm. Since the volume has already been defined, the radius can be expressed as a diameter, which has also been defined, so all that remains is to calculate this distance.

$$A = \pi r^2 \tag{1}$$

$$V = h * A \tag{2}$$

Where "r" represents the radius of the pipe in which the LEDs are installed, "A" the area of the circumference of the pipe, "h" the distance the pairs of diodes are from each other and "V" the volume of the disinfectant fluid for one dosage.

Figures 21 and 22 show, respectively, the diagram of the electrical circuit used for each diode pair generated in Easyeda and the chlorine level sensor developed for this book.

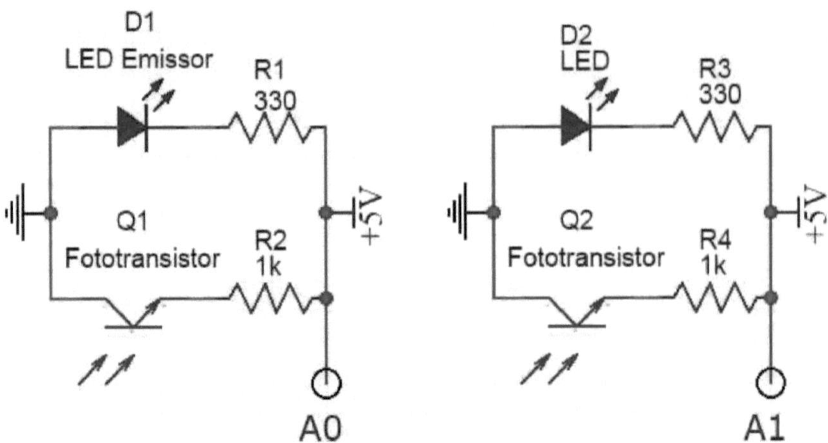

Figure 21 - Schematic drawing of the chlorine level sensor electrical circuit.
Source: prepared by the author

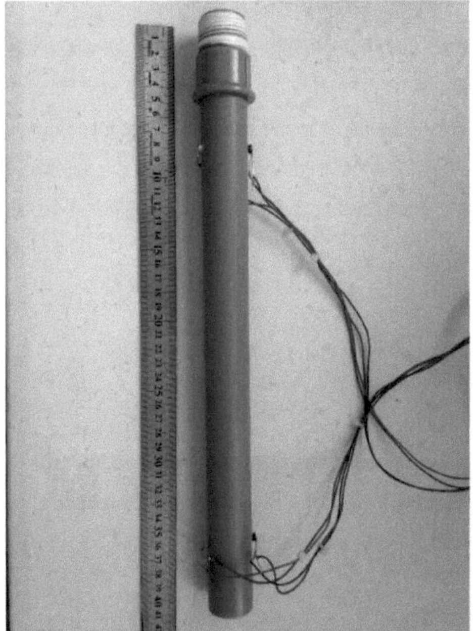

Figure 22 - Chlorine level sensor.
Source: prepared by the author

3.1.2 Rainwater level sensor.

This sensor was developed using the same operating principle as a metal electrode level sensor, as it would be easy to include and develop this type of device in the project, as well as having a relatively low manufacturing cost.

The sensor required 7 stainless steel 1/8 inch bolts and nuts, 7 *o-rings to prevent* water infiltration, a 1/2 inch PVC (Polyvinyl Chloride) knee, 6 1/2 inch PVC **"T" fittings,** 7 1/2 inch PVC caps, all 1/2 inch PVC and, finally, electrical wiring. Figure 23 shows the rainwater sensor developed for this book.

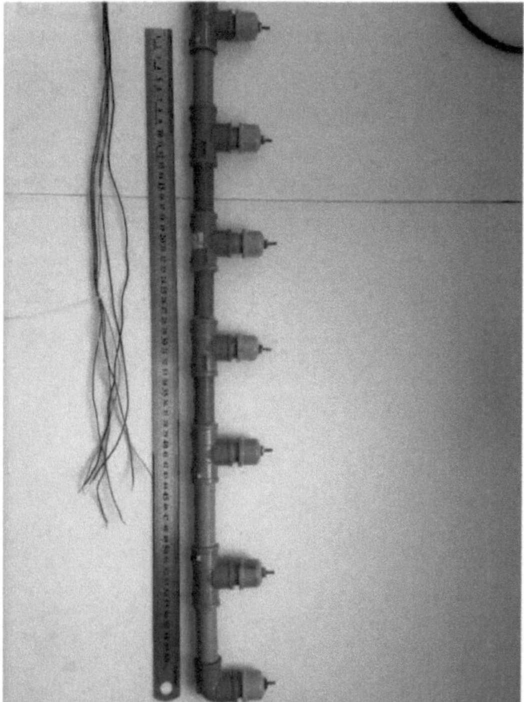

Figure 23 - Rainwater level sensor.
Source: prepared by the author

Before starting to build the sensor, we checked how many digital ports the Arduino UNO had available, after installing the chlorine level sensor, display and peristaltic pump drive relay, obtaining the result of 6 ports, counting one more for GND seven, so the device for this case could only indicate six levels of discrete rainwater measurements.

A water tank with a storage capacity of 1000 litres was taken as the basis for the book, as these are often used in homes. Therefore, by subdividing the seven electrodes into equal parts, our device is able to measure approximately 166 litres of water per rod interval.

The device has a lower reference metal rod connected to the Vcc of the power supply circuit, and 6 more discrete level indicator metal rods, each connected respectively to its signal transducer transistor base, which acts like a switch.

When the sensor is triggered, the reading signal, which is binary, is made between the transistor's emitter and a *pull-down* resistor, which guarantees a zero reference logic level when the transistor is not conducting.

When the water level exceeds the lower metal rod or any other measuring rod, the circuit between the electrodes is closed and a base current flows through the signal transistor, closing the channel between the transistor's collector and emitter. Using the controller developed for the system, it is possible to read the signal relating to the level of the reservoir in question.

Figure 24 shows the typical electrical circuit for an electrode pair created by Easyeda, used in this book. The image shows the type of transistor used in the project as well as the resistor.

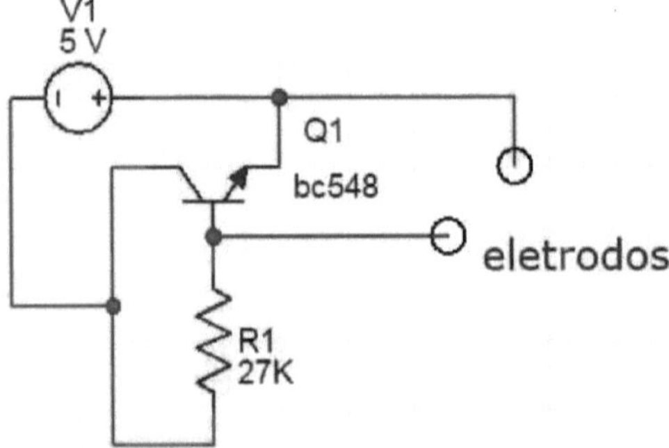

Figure 24 - Drawing of the electrical circuit of a pair of electrodes.
Source: prepared by the author.

3.3 Controller.

The Arduino UNO was used as the controller. An electronic prototyping platform with free, single-board hardware, designed with an Atmel AVR microcontroller and a programming language that is essentially C++. This component is seen as the core of the project, as all the control and management involved depends on signals sent or received by it. Figure 25 shows a model of an Arduino board.

Figure 25 - Arduino board model.
Source: www.filipeflop.com

Analogue and digital ports were used in this book. The A0 and A1 analogue ports were used to read the infrared radiation receiver diodes and the chlorine sensor. Part of the digital ports were assigned to the rainwater sensor; each electrode is associated with one of these ports, the other ports with the display and the peristaltic pump drive relay.

The Arduino controller was connected to a computer via a USB AM/BM cable and, via the IDE interface, the device was programmed with an algorithm in the C++ language. All the components were interconnected using conductive wires on a protoboard.

3.4 Peristaltic pump.

The bleach is dosed by activating the peristaltic pump via the 5-volt relay when the controller sends the activation information. Figure 26 shows the model of the pump used in the book.

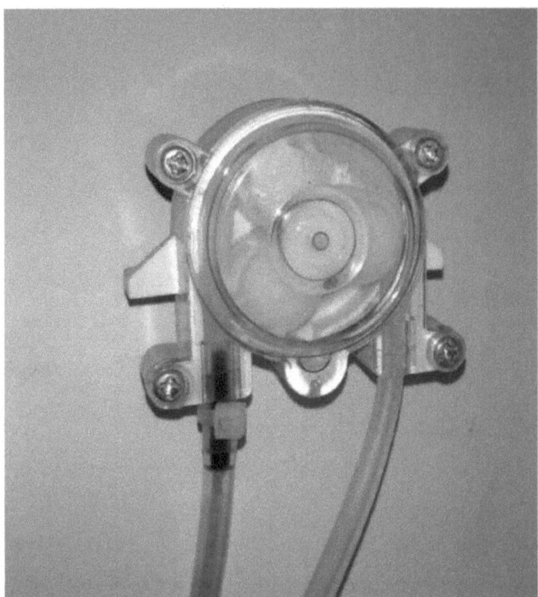
Figure 26 - 12 volt peristaltic pump.
Source: prepared by the author

This type of pump is often used in chemical analysis laboratories and fish tanks. This type of pump is suitable for the project, because as the working fluid does not come into contact with the internal parts of the pump, only with the fluid transport hose, there is no possibility of oxidation and consequently premature damage to the internal parts. This is because the fluid in question is considered corrosive.

Considering that the portion of this liquid dosed is small, only 166ml of disinfectant per discrete measurement interval of the rainwater level sensor, the flow rate of this volume, matched by this peristaltic pump operating at 12 volts, proved to be adequate for the project, since the transfer of the disinfectant fluid from its reservoir to its final destination, the rainwater storage tank, takes place in a short time interval.

Among these characteristics. This equipment could also be easily incorporated into the automated closed-loop system developed in this book, and is also sold at a low cost on the market.

The relay works with an electromechanical switch. The physical displacement of this switch occurs when the electric current travels through the coils of the relay, thus creating a magnetic field which in turn attracts the lever responsible for changing the state of the contacts. The relay was used because it is the easiest and simplest way to activate the peristaltic pump with a 5 volt electrical pulse. Figure 27 shows a 5-volt relay similar to the one used in the book.

Figure 27 - 5v relay.
Source: prepared by the author

3.5 Display.

A display is a thin panel used to display information electronically, such as text, images and videos. In the project, a display (16x2) was used, on which two lines of text can be displayed. Using the information sent from the controller to the device, messages are sent to the operator about the level of rainwater in the storage tank, sanitary water level alerts, the start of disinfection and preliminary messages about the initialisation of the prototype. Figure 28 shows a display model similar to the one used in the book.

Figure 28 - 16x2 display.
Source: prepared by the author

When the system is activated, the Arduino sends a digital signal to the display to show the operator the preliminary messages, just once. This information corresponds to a sequence of sentences about initialising the prototype in the following order:

- Sara Project
- Check for disinfectant
- Feeder started

At the end of this preliminary information expressed by the display, the operator will be presented with constantly updated messages about the amount of bleach dosed, the need to refill the disinfectant, the last dosage, the discrete level of the rainwater and the start of the disinfection process. Figure 29 shows the messages sent by the rainwater and chlorine level sensors. In a) you see the message from the first disinfection and the volume of sanitary water injected into the rainwater reservoir in b) you have the same message from the first disinfection and the discrete level of rainwater in the storage reservoir, and figure 30 shows the messages coming from the chlorine level sensor. In a) you see the message referring to the last dosage, in b) you see the message about the replenishment of disinfectant in the chlorine tank.

a)   b)

Figure 29 - First disinfection and quantity of disinfectant and rainwater.
Source: prepared by the author

a)   b)

Figure 30 - Last dosage and refill with disinfectant.
Source: prepared by the author

3.6 System assembly.

The set-up for operating and calibrating the prototype developed in this book is shown in figure 31 below. The structure has a wooden bench, where you can see the connected devices, the rainwater and sanitary water tanks, the level hose for visualising the level of stored rainwater and the drain tap for emptying the rainwater tank.

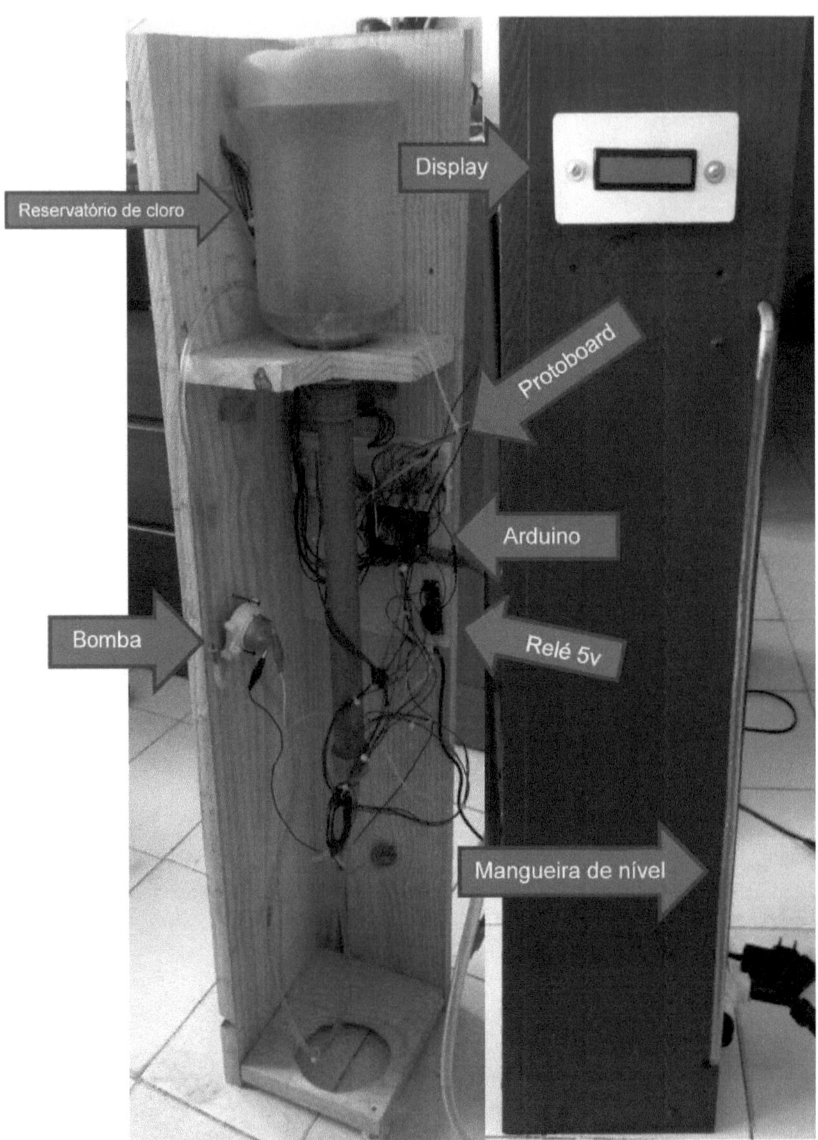

Figure 31 - Assembly of the prototype.
Source: prepared by the author

The controller, display, protoboard, relay and peristaltic pump are fixed to the prototype's demonstration structure, while the rainwater level sensor **is submerged in the tank, which simulates a** 1000 litre **residential water tank.** The chlorine level sensor is supported by a bracket on the structure, with the upper side connected to the disinfectant tank and the lower side connected to the peristaltic pump suction hose, while the bleach injection hose directs the disinfectant to the rainwater tank.

CHAPTER 4

Results and Discussions
4.1 Calibration of the chlorine level sensor.

This sensor was calibrated in order to see if it was operating as intended. The calibration test used a PVC pipe with the same dimensions and characteristics as the secondary reservoir. Figure 32 shows the test bench on which the chlorine level sensor was tested.

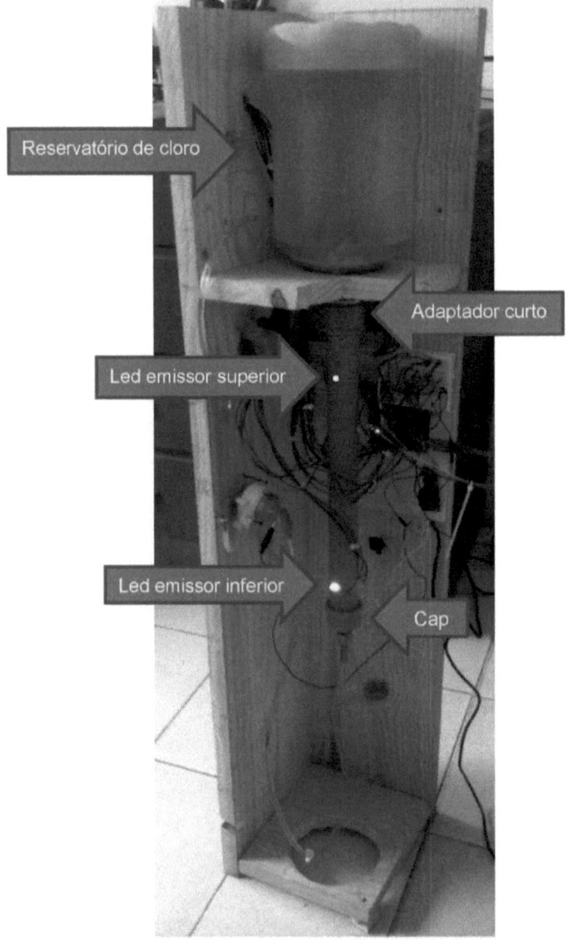

Figure 32 - Chlorine level sensor test bench.

Source: prepared by the author

Water was used for the test, filling the secondary reservoir with the aid of a funnel and emptying it with the suction hose connected to the peristaltic pump. It was observed that when the water level submerged the IR diode and transistor, values between 939 and 967 were displayed. Figure 33 shows the values generated on the Serial monitor in the presence of liquid.

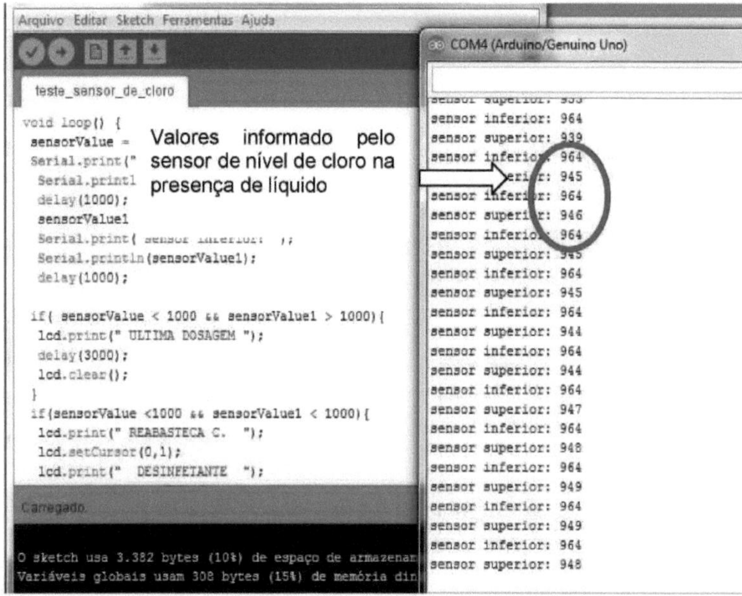

Figure 33 - Result on the serial monitor in the presence of liquid.
Source: prepared by the author

While in the absence of liquid, values between a range of different numbers 1009 to 1018 were observed. Figure 34 shows the values displayed by the serial monitor when the secondary reservoir was empty.

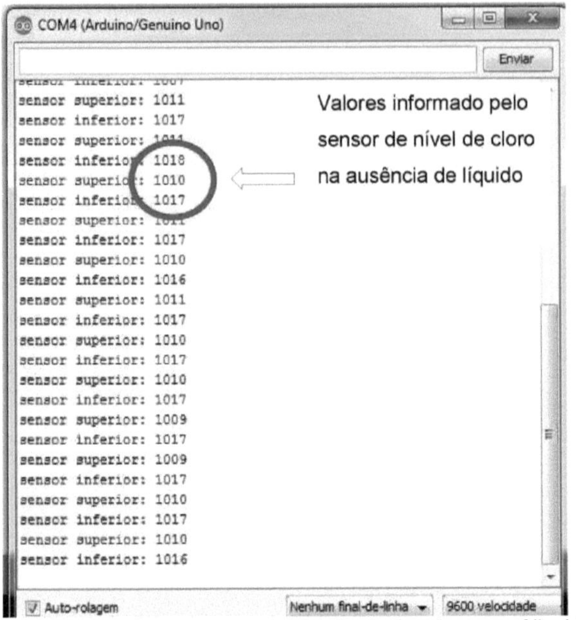

Figure 34 - Results on the serial monitor in the absence of liquid.
Source: prepared by the author

By observing the behaviour of the level sensors based on receiving infrared rays (see Figures 33 and 34), it was assumed that when the Arduino received values greater than 1000, there was a probability that the reservoir was empty, while values below 1000 meant that the reservoir was more likely to be full.

4.2 Testing the rainwater level sensor.

The rainwater level sensor test was carried out using a reservoir, **simulating a 1000L water tank,** made from PVC with a diameter of 100mm, a 100mm cap attached to the bottom of the pipe, a drain valve also installed at the bottom of the reservoir and a level hose to make it easier to see the volume of water inserted into the reservoir. Figure 35 shows the tank made for the test.

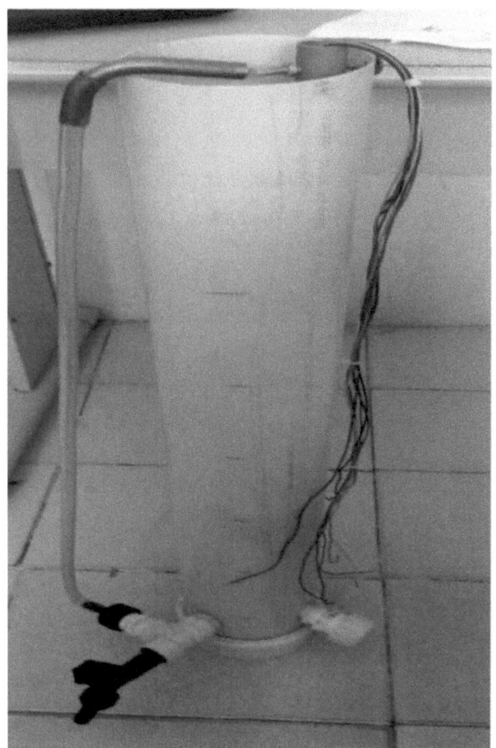

Figure 35 - Rainwater level sensor test tank.
Source: prepared by the author

The level sensor was fixed to the tank with a bolt and nut so that the device in question would remain static, as movement up or down of the sensor is a source of error.

With a Becker in hand, water was added and the volume of liquid required was counted until the sensor indicated its electrical stimulus to the Arduino, which was displayed via the Serial monitor. A value of zero was displayed when the water did not reach the electrode in question and a value of 1 when the water level submerged the electrode. For each electrode measurement level, five tests were carried out to gauge the volume of fluid. Table 4 shows the test results.

Table 4 - Rainwater level sensor test results.

Sensor level	Average volume (L)	Average difference between each electrode (L)	Average volume measured by the sensor
1	0,84	0,84	
2	1,72	0,88	0,84
3	2,56	0,84	
4	3,37	0,81	

| 5 | 4,2 | 0,83 |
| 6 | 5,05 | 0,85 |

Source: prepared by the author

Observing the results shown in Table 5 and using a simple rule-of-three calculation, it was found that on average each electrode interval of the sensor made measures a discrete volume of approximately 166.67 litres of rainwater when installed in a 1000 litre water tank.

4.3 Peristaltic pump flow test.

In order to determine the flow rate of the peristaltic pump, a source of varying voltage, a container graduated in ml and a small reservoir containing water were used. Figure 36 shows a diagram of how the flow rate test was carried out.

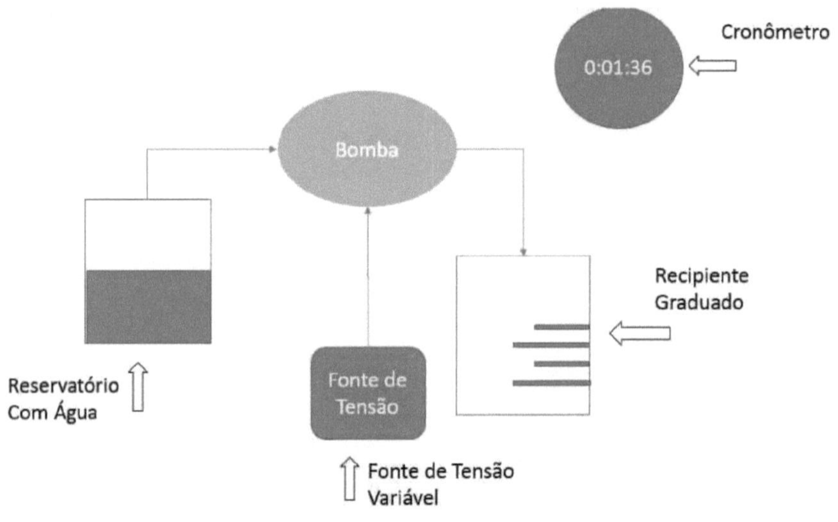

Figure 36 - Flow test diagram.
Source: prepared by the author

When the peristaltic pump was energised, the hose submerged in the water tank extracted the fluid and directed it to the graduated container. With a stopwatch in hand, the time interval in minutes taken to fill the 100 ml volume of the graduated container was measured. The pump's drive voltage started at 6V and for each experiment 1V was added until it totalled 12V. For each voltage value, 5 tests were carried out in order to find an average value for the time taken to fill the 100ml volume for each voltage value. To protect the pump, no tests were carried out with voltages above 12V. Table 5 shows the average time for each voltage test carried out.

Table 5 - Measurement of the peristaltic pump flow test.

Voltage (V)	average time	Volume (ml)	Flow

	(min)		(ml/min)
&	2,30	100	43,48
7	1,64	100	60,90
8	1,26	100	79,37
9	1,05	100	95,06
10	0,90	100	111,11
11	0,80	100	125,00
12	0,72	100	138,89

Source: prepared by the author.

The flow rate was obtained by dividing the determined volume of 100 ml by the time taken for the peristaltic pump to reach this capacity, according to equation 3.

$$Q = \frac{V}{t}$$
(3)

Where Q is the flow rate in ml/min, V is the volume in ml, t is the time in seconds.

The graph in figure 7 shows a comparison between the flow behaviour of the peristaltic pump for different voltages and a linear regression.

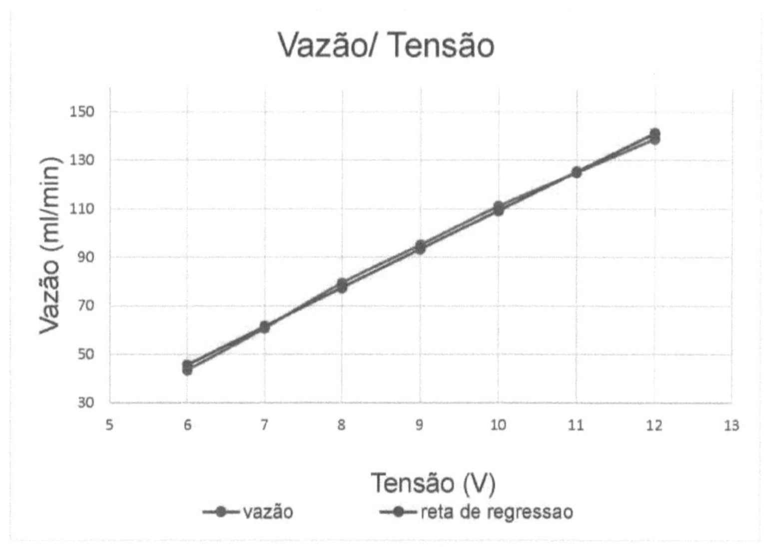

Figure 37 - Comparison of the flow rate test and linear regression.

Source: prepared by the author

As can be seen from this graph (figure 37), the flow rate increased proportionally with the increase in voltage. Using the regression line shown above as a reference, we can define the time interval needed to run the peristaltic pump to dose any desired volume of liquid.In view of the results of the flow rate tests shown in Table 4, the voltage used to drive the pump in this book was 12 V, as it has a higher flow rate, showing a shorter time interval for dosing.As each electrode interval of the rainwater level sensor measures 166L of rainwater, a volume of 166ml of bleach is required for each disinfection. Therefore, from equation 4 extracted from the graph above (figure 37) and equation 3, it is decided that the operating time of the peristaltic pump for each disinfection is 1.18 minutes.

$$Y = 15{,}93X - 50{,}01 \qquad\qquad (4)$$

This is a first degree polynomial equation, where X represents the value of the voltage adopted, Y the value of the flow rate obtained, with an angular coefficient of 15.93 and a linear coefficient of -50.01.

During the flow test, it was noted that after 20 minutes of continuous operation, the pump showed an increase in temperature. In order to avoid premature damage to the equipment, it is suggested that when this time interval is reached in uninterrupted operation, the device should be deactivated for a certain time until it reaches an ambient temperature as a form of prevention.After determining the time interval sufficient to carry out the dosage for disinfecting the rainwater using equations 3 and 4, it was ascertained whether this time of 1.18 minutes was really necessary to inject 166ml of bleach through the pump. Five tests were carried out to determine the average volume in ml of fluid being injected by the 12 volt peristaltic pump. Table 6 shows the results obtained in each test.

Table 6 - Peristaltic pump injection volume test.

Essay	Time (min)	Volume (ml)	Average (ml)
1	1.18	169	
2	1.18	166	
3	1.18	165	167,2
4	1.18	169	
5	1.18	167	

Source: prepared by the author.

After observing the results, it was found that the time interval adopted for operating the pump with the aim of disinfecting it was very satisfactory, as it showed a very small percentage error of just 0.72%.

4.4Final test.

Finally, all the prototype devices were simulated together to form the automated system proposed in the book, in order to assess their performance.

For this final test, all the devices that make up the automated system were assembled on the aforementioned wooden bench. Figure 38 shows all the items fitted together.

Figure - 38 Final test.
Source: prepared by the author.

During its operation, the prototype executed all the commands synchronously, as proposed in this book. The two sensors developed worked perfectly as described above. There was no unexpected behaviour during operation, so the closed-loop automated system developed proved worthy of being used for rainwater disinfection.

CHAPTER 5

Conclusions

Given the drastic growth in demand for water in all sectors of production, natural factors and sometimes inefficient public management of this water resource have contributed to certain regions suffering from its scarcity, while others have faced this problem in recent years.

In view of these aspects, the search for alternative methods of supplying drinking water is indispensable in cases where an eminent degree of potability is not required, but the application of an appropriate treatment is essential.

There are various ways of harnessing, reusing and disinfecting water. One of these is the utilisation of rainwater. A catchment system can be installed in a home and this resource, once stored, can be disinfected with bleach in appropriate concentrations in order to be used for non-potable purposes such as watering the garden, toilet bowls and so on.

This book aims to do just that. To disinfect rainwater, previously collected by a catchment system, with sanitised water in appropriate proportions, using an automated system, in order to prevent diseases transmitted by contact with contaminated water. This was possible after the development of two different sensors, which worked in conjunction with the Arduino UNO microcontroller, a peristaltic bucket and an LCD display, creating a low-cost automated disinfection system.

When designing the prototype, the focus was on using bleach as a chemical disinfectant, since it can be easily purchased and found anywhere at a lower cost than other chemical disinfectant compounds.

In short, it is feasible to disinfect rainwater for non-potable purposes. This action contributes to the eradication of water-borne diseases. Furthermore, the prototype developed demonstrates the feasibility of building an efficient, easy-to-operate, automated system with little human interaction, capable of disinfecting rainwater properly.

References

ABES - ASSOCIAÇÃO BRASILEIRA DE ENGENHARIA SANITÁRIA E AMBIENTAL Revista Engenharia Sanitária e Ambiental, v. 2, n.° 3 jul/set/97 Rio de Janeiro, 1997.

ACCARDI, Adonis; DODONOV, Eugeni. **Home Automation: Basic Elements, Architectures, Sectors, Applications and Protocols.** In: Tecnologias, Infraestrutura e Software,

São Carlos, v. 1, n. 2, p. 156-166, nov. 2012. (ISSN 23162872).

AIETA, M. AND ROBERTS, P.V. **Determination of chlorine dioxide, chlorite, and chlorate in water**. Journal AWWA, Research, and Technology, p. 64-69, January, 1984.

ANA, FIESP and SindusCon-SP. Conservation and Reuse of Water in Buildings. São Paulo: Prol Editora Gráfica. 2005.

ANNECCHINI, K. P. V. Harnessing **rainwater for non-potable purposes in the metropolitan region of Vitória (ES)**. 2005. 124 p. Dissertation (Master's in Environmental Engineering) - Federal University of Espírito Santo, Espírito Santo.

BARCELLOS, B. R.; FELIZZATO, M. R.; Utilisation of atmospheric water. In: 23rd BRAZILIAN CONGRESS OF SANITARY AND ENVIRONMENTAL ENGINEERING, 2005, Campo Grande. Brazilian Environmental Sanitation: Utopia or Reality? Campo Grande: ABES. p. 112 - 112. CD-ROM. 2005.

BEGA, Egídio Alberto et al. **Instrumentação Idustrial**. 2. ed. Rio de Janeiro: Interciência, 2006. 583 p.

CARVALHO, Gabriela dos S. **Analysis of a proposal for a rainwater collection and utilisation system for use in toilet bowls in single-family homes**. Rio Claro, SP: 2007. 64 f. Monograph. Undergraduate course in Environmental Engineering at the Institute of Geosciences and Exact Sciences - Unesp, Rio Claro Campus (SP).

CAVINATTO, V. M. **Saneamento básico: fonte de saúde e bem-estar**. São Paulo: Ed. Moderna, 1992.

CHANG, S. Morris, J. Elemental iodine as a disinfectant for drinking water. **Industrial and Engineering Chemistry**. 1953; 45:1009.

EPA - Environmental Protection Agency. Ozone disinfection. EPA 832-F-99- 063. US EPA, Washington DC, Sep. 1999a.

GOTTARDI W. Iodine and iodine compounds. In: Disinfection, Sterilisation, and Preservation (Block S, ed). **Philadelphia:Lea & Febiger**, 1991;152-167.

GRASSI, M. T.; JARDIM, W. F. Ozonisation of water: chemical and toxicological aspects. Revista DAE - SABESP, v.53, n.º 173, p. 01-06, Sep/Oct 1993.

GROUP RAINDROPS. Harnessing Rainwater. Curitiba, 2002.

Home Use of Greywater, Rainwater Conserves Water - **and May Save Some Money**. Available at < https://wrrc.arizona.edu/publications/arroyo- newsletter/home-use- 60 greywater-rainwater-conserves-water-and-may-save- money>. Accessed on:11 Nov. 2016.

LAUBUSCH, E. J., 1971. Chlorination and other disinfection processes. In: *Water Quality and Treatment:* **A Handbook of Public Water Supplies (American Water Works Association)**, pp. 158-224, New York: McGraw-Hill Book Company.

LEMOS, Diego Araújo. **Automatic rainwater collection and disposal system.** 2016. 55 f. TCC (Graduation) - Mechanical Engineering Course, Technology Centre, Federal University of Rio Grande do Norte, Natal, 2016.

LESCHBER, R.; PERNAK, K. D.; ZIMMERMANN, U. Investigation of the behaviour of inorganic and organic substances during infiltration of rain runoff (in German). **Stadtentwasserung und Gewasserschutz**, 4, p. 169-188, 1991.

MANO, R. S.; SCHMITT, C. M. Captação Residencial de Água Pluvial, para Finsos Não Potáveis, em Porto Alegre: Aspectos Básicos da Viabilidade Técnica e dos **Benefícios do Sistema. CLACS' 04 -** I Latin American Conference on Sustainable Construction and ENTAC 04, - 10th National Meeting on Built Environment Technology, São Paulo - SP, Anais....CD Rom, 2004.

MAY, S; PRADO, R.T.A. **Estudo da viabilidade do aproveitamento de água de chuva para consumo não potável em edificações.** São Paulo, SP:

2004. 190 f. Master's dissertation. Postgraduate Programme in Civil Construction Engineering, Polytechnic School, University of São Paulo.

MAY, Simone. **Characterisation, Treatment and Reuse of Greywater and Use of Rainwater in Buildings.** Thesis (Doctorate in Engineering) - Department of Hydraulic and Sanitary Engineering, Polytechnic School of the University of São Paulo, São Paulo, 2009. 223p.

MEDEIROS, C. O.; LIMA, R. R.; MARTINS, R. A.; BUENO, K. L.; DIEL, J. V.; RODRIGUES, L. M.; SOUZA, T. R.; "STUDY OF SEDIMENTATION FOR THE TREATMENT OF PROCESS WATER FROM A RICE PROCESSING INDUSTRY", p. 9458-9463 . In: **Proceedings of the XX Brazilian Congress of Chemical Engineering - COBEQ 2014 [= Blucher Chemical Engineering Proceedings, v.1, n.2].** SãoPaulo : Blucher, 2015.
ISSN 2359-1757, DOI 10.5151/chemeng-cobeq2014-1935-16796-133460.

MEYER, S. T.; The use of chlorine in water disinfection, the formation of trihalomethanes and the potential risks to public health. **Cad. Saúde Pública**, Rio de Janeiro 10 (1): p. 99-110, Jan/Mar, 1994.

MIERZWA, JOSÉ CARLOS; HESPANHOL, IVANILDO, **Water in Industry: Rational Use and Reuse.** Oficina de Textos - São Paulo, 2006. 143p.

MIURA, J., SILVA, A. B. H., LIMA, J. D. C. V. "Harnessing rainwater for non-potable purposes in an industrial development in the centre-west of Brazil". Paper from the XIII Ibero-American Symposium on Water, Sewage and Drainage Networks, Fortaleza - CE, November 2014.

NAKADA, Liane Yuri Kondo. **TREATMENT OF RAINWATER FOR NON-POTABLE PURPOSES USING CORN STARCH AS A COAGULANT IN CYCLIC FILTRATION ON A LABORATORY SCALE.** 2008. 35 f. TCC (Graduation) - Environmental Engineering Course, Institute of Geosciences and Exact Sciences, Paulista State University, Rio Claro, 2008. Available at: <file:///C:/Users/Stone/Desktop/TCC_escrito/nakada_lyk_tcc_rcla.pdf>. Accessed on: 18 Nov. 2016.

WHO - WORLD HEALTH ORGANISATION. Water disinfection. Washington DC, 1999.

UN, 2014, The volume of fresh water resources on Earth is around 35 million km3. Available at:<http://www.unwater.org/statistics/statistics-detail/en/c/211801/>. Accessed on 11 September 2016.

PEREIRA, Fábio Sérgio da Costa. **Water reuse through alternative sources.** 2014. Available at: <http://www.crea-rn.org.br/artigos>. Accessed on: 8 October 2016.

Philippi Jr A. (ed.). Sanitation, Health and the Environment. Barueri, SP: Manole; 2005.

PINHEIRO, José Mauricio Santos. **Automation Systems.** 2004. Available at: <http://www.projetoderedes.com.br>. Accessed on: 14 Nov. 2016.

PINHEIRO, Pedro. **Water-borne diseases.** 2016. Available at: <http://www.mdsaude.com/2012/01/doencas-da-agua.html>. Accessed on: 10 Nov. 2016

PINHO, Luzimar de Souto Amorim Ribeiro. **LEPTOSPIROSIS.** 2002. Available at: <http://www.rondonia.ro.gov.br/agevisa/institucional/vigilancia-ambiental/controle-de-zoonoses-e-animais-peconhentos/leptospirosis/>. Accessed on: 16 October 2016.

PORDEUS, Arthur Nobre. **Automatic washing machine water reuse system.** 2016. 74 f. TCC (Graduation) - Mechanical Engineering Course, Technology Centre, Federal University of Rio Grande do Norte, Natal, 2016.

R. C. FUENTES. Industrial automation workbook. Federal University of Santa Maria, Santa Maria Industrial Technical College, 2005.

RIBEIRO, A. C; ALVAREZ, V. H.; GUIMARÃES, P. T. G. Comissão de fertilidade do solo de Minas Gerais. Recomendações para o uso de corretivos e fertilizantes em Minas Gerais - 5° Aproximação, Viçosa, MG, 1999 p.359.

RIBEIRO, Marco Antônio. **Industrial automation.** 4. ed. Salvador: Tek, 2001.

ROSÁRIO, João Mauricio. **Industrial automation.** São Paulo: Baraúna, 2009. 514 p.

SIGHIERI, L. Chlorine. Properties, production, handling, hazards, safety conditions. Water Disinfection, Chap. 5, p. 29-46.CETESB, São Paulo, 1974.

SOUZA, M. A. A. Solar disinfection: proposal for a feasibility study methodology and determination of basic parameters. Proceedings of the 20th Brazilian Congress of Sanitary and Environmental Engineering - Rio de Janeiro. RJ. 2000. CDROOM.

TAVARES, A. C. **Physical, chemical and microbiological aspects of water stored in cisterns in rural communities in the semi-arid region of Paraiba**. Master's dissertation, Federal University of Campina Grande, Campina Grande, PB, 2009.

Thayer, A.; Chem. Eng. News 2007, 85 (25), 70

Tomaz, P. Utilisation of rainwater in urban areas for non-potable purposes, 2010. Available at: < http://pliniotomaz.com.br/livros- digital/>.Accessed on 12 November 2016.

TUCCI, Carlos Eduardo Morelli. **World Water Day: increasing demand and contamination cause concern.** 2013. Available at: <https://noticias.terra.com.br/ciencia/sustentabilidade/dia-mundial-da-agua-aumento- da-demanda-e- contaminacaopreocupam,00c5affa3719d310VgnVCM4000009bcceb0aRCRD.html>. Accessed on: 3 October 2016.

ZILLICH, G. Requirements for rainwater in Lower Saxony (in German). Berichte der Abwassertechnischen Vereinigung, 41, p. 609-613, 1991.

ANNEX A - budget with the main components.

Cost of building the prototype	
Description	**Value**
1 arduino UNO	R$ 16.00
1 peristaltic pump	R$ 70.00
1 protoboard	R$ 15.00
1 5V relay	R$ 11.00
1 16x2 display	R$ 5.00
1 x 2.1 litre plastic container	R$ 9.90
1 cap **1"**	R$ 2.20
1 m of pipe 1 "	R$ 8.30
1 weldable adapter **with ring for water tanks**	R$ 6.90
1 weldable short adapter	R$ 6.90
2 led emitter	R$ 0.60
2 phototransistor	R$ 1.00
2 resistors 330 Q	R$ 0.20
2 resistors 1K Q	R$ 0.20
1 metre of *Vz* **pipe**	R$ 3.15

7 cap *Vz*	R$ 1.50
6 14" tee fittings	R$ 8.95
1 weldable knee 14"	R$ 1.00
12 metres of wire	R$ 6.00
7 transistors	R$ 1.05
6 resistors 27k Q	R$ 0.60
7 stainless steel bolt and nut 1/8	R$ 1.05
TOTAL	**R$ 176.50**

I want morebooks!

Buy your books fast and straightforward online - at one of world's fastest growing online book stores! Environmentally sound due to Print-on-Demand technologies.

Buy your books online at
www.morebooks.shop

Kaufen Sie Ihre Bücher schnell und unkompliziert online – auf einer der am schnellsten wachsenden Buchhandelsplattformen weltweit! Dank Print-On-Demand umwelt- und ressourcenschonend produzi ert.

Bücher schneller online kaufen
www.morebooks.shop

info@omniscriptum.com
www.omniscriptum.com

Printed by Books on Demand GmbH, Norderstedt / Germany